MAKE MATH YOUR PATH

FRAGMATICS

A SELF EDUCATION BOOK ON EVERY FRAGMENT OF VEDIC MATHEMATICS

Akshatha Amruth

Made with ❤ on the Notion Press Platform

www.notionpress.com

Contents

Preface

Dear reader,

It gives me great pleasure to present to you this comprehensive guide to Vedic Mathematics. In the pages that follow, you will embark on a journey through ancient India's mathematical insights and witness the astounding power of Vedic techniques.

The origin of Vedic Mathematics can be traced back to the ancient texts of the Vedas, which not only encompass profound spiritual wisdom but also harbour a treasure trove of mathematical knowledge. These ancient sages realized the beauty and simplicity of mathematics and sought to transmit their wisdom through concise techniques that can be easily learned and applied.

This book aims to introduce you to the elegance of Vedic Mathematics, taking you step by step through various principles, concepts, and methods. Whether you are a student, a teacher, or simply someone with a curious mind, this book will equip you with techniques to solve complex mathematical operations swiftly and effortlessly.

One of the most remarkable aspects of Vedic Mathematics is its ability to simplify calculations by eliminating tedious and time-consuming steps. By utilizing sutras or formulae, Vedic Mathematics offers intuitive shortcuts that enable you to solve problems mentally and with lightning speed. The simplicity and efficiency of these techniques are truly remarkable, providing a refreshing alternative to conventional mathematical methods.

Each chapter is carefully structured with clear explanations, illustrative examples, and practice exercises to ensure a comprehensive grasp of the subject matter.

It is important to note that Vedic Mathematics is not limited to computation alone. It emphasizes a holistic approach, encouraging deep mathematical understanding and fostering creativity. By exploring the underlying principles and underlying patterns, Vedic Mathematics cultivates analytical thinking skills, making it a valuable tool for problem-solving and logical reasoning in various disciplines.

As you delve into this book, I encourage you to embrace the spirit of exploration and immerse yourself in the world of Vedic Mathematics. Discover the joy in unravelling mathematical complexities and witness the transformation of your own mathematical abilities.

In conclusion, I hope this book serves as a guide to unlock the wonders of Vedic Mathematics and ignite a lifelong fascination with the ancient wisdom of the Vedas.

Happy reading!

Sincerely,

Akshatha Amruth

Acknowledgements

1. I thank Smt. M V Vasavi, my teacher, guru and mentor in the field of Vedic Mathematics. She has taught me and inspired me in many ways to further my pursuance in the field of Vedic Mathematics.

2. I thank my husband Amruth A Makam who has been my constant support throughout this journey.

3. I would like to thank my family for their support, encouragement, and understanding during the writing process.

4. I thank all my students who have inspired me. They have shown me that the subject I teach matters and their eternal interest to learn more has encouraged me to write this book and spread the knowledge of Vedic Mathematics.

5. I express my gratitude to Notion Press who have helped me in the publication of the Book.

About The Author

Akshatha Amruth is an expert in the field of Vedic Mathematics. She has dedicated her career to studying and teaching Vedic Mathematics, inspiring a number of individuals to explore its principles and techniques.

Akshatha holds an MTech in Structural Engineering, providing her with a solid academic foundation to delve into the intricacies of Vedic Mathematics and its applications in the real world. She wishes to make her expertise accessible to a wider audience and hence has authored this book.

Through her engaging writing style and ability to simplify complex concepts, Akshatha has earned a reputation for being an exceptional educator. She strives to make learning mathematics a joyful and enlightening experience for students of all ages and backgrounds.

In addition to her academic pursuits, Akshatha actively conducts workshops and seminars on Vedic Mathematics, sharing her knowledge and insights with eager learners around the world. She is deeply committed to preserving and promoting the wisdom of the Vedas, believing in the transformative power of Vedic Mathematics in our modern lives.

With her profound understanding of Vedic Mathematics and dedication to spreading its teachings, Akshatha continues to inspire and empower individuals on their mathematical journey."

To embark on your journey in the field of Vedic Mathematics, reach out to her.

Email id: fragmatics.vedicmath@gmail.com

Introduction to Vedic Mathematics

Each one of you must have come across the word "Vedic Mathematics" but would be wondering what exactly this is and how it works.

Vedic Mathematics is not a separate branch of mathematics but, instead it's a name given to ancient system of calculation which was rediscovered from the Vedas specially from the **Atharva Veda** between 1911 and 1918 by **Sri Barathi Krishna Tirthaji Maharaj**. According to him the whole of Vedic Mathematics is based on the **"16 sutras" and "13 sub-sutras"**.

Using these sutras any part of mathematics such as addition, multiplication, division, subtraction, squares, square roots, cube, cube roots, algebra etc. can be easily and quickly solved.

Learning Vedic Mathematics makes mathematics fun and enjoyable to the learners. It enhances the speed of calculation without using a calculator thus saving lot of time during competitive examinations. The most complicated problems can be solved very quickly there by, enhancing and widening the intelligence of the learners by making them love mathematics.

Chapter - 1

Beejank

This is used for quick verification of problems. The word Beejank is derived from Sanskrit in which Beej meaning one or single and ank meaning number. Together meaning single digit number. So Beejank of a number means the single digit of any given number. This is obtained by adding all the digits in the given number. In case the sum of digits turns out to be more than 1 digit, then these digits are in turn added till we get a single digit number (Beejank).

This sutra is used in verifying our answers.

To find the Beejank of a number we adopt two rules which are as follows:

- Strike out the number 9 in the series of numbers.
- Strike of the numbers when added gives a sum of 9.

Note: Suppose all the numbers gets striked out in the series and there is no number left behind we write the beejank of that number as 9.

Let us have a look at some numbers as to how to find Beejank.

Example -1: Find the Beejank of 27365

In this number series 27365 first we should check for number 9 since it's not there now let's try to find out the numbers when added gives a sum of 9.

So (6+3=9) and (7+2=9) so let's strike off 7,2,6,3 we are left out with number 5 which is a single digit number.

So, the **Beejank is 5.**

Example - 2: Find the Beejank of 4069

In this number series we have 9 so let's strike off 9.

Now, we are left out with 4,0,6. By adding all the 3 numbers 4+0+6 we get 10 which is a 2-digit number so we again add 1+0 which gives 1.

So, the **Beejank is 1.**

Example – 3: Find the Beejank of 27111429

We have 9 so let's strike off the number 9. When we add (7+2) we get 9 so let's strike out 7 and 2. Now, when we add (4+2+1+1+1) we get 9 so let's strike out 4,2,1,1,1. We are not left out with any number behind.

So, the **Beejank is 9.**

A few more examples are given below:

759 = 3

26237 = 2

546372 = 9

77 = 5

4215783005 = 8

Exercise problems:

1. 230
2. 698204
3. 7116219
4. 6543726
5. 9945297643598

6. 974521639
7. 438
8. 4401
9. 82617239721
10. 5896457

RIDDLE – 1: Do you know the other name of a circle?

Chapter - 2

Nikhilam, Ekhadhikena and Ekanyunena Purvena Sutras

Nikhilam Navatascarama Dasatah sutra is used for easy subtraction. This sutra means **"All from 9 and last from 10"**.

This sutra is used to find the complement of a number in actual mathematics.

According to this sutra the last number (unit digit) has to be subtracted from 10 and the remaining numbers have to be subtracted from 9.

We should always start subtracting from the last number.

Note: While applying Nikhilam sutra if the last number is zero then, we should write zero as it is and the number next to zero becomes the last number.

Ekhadhikena means the next number or (successor). Ekh means 1, Adhika means more so 1 number more.

Ekhanyunena Purvena means the before number or 1 number less or (predecessor).

Now let's try out some examples:

Numbers	Nikhilam	Ekhadhikena	Ekanyunena
627	373	628	626
490698	509302	490699	490697
8972	1028	8973	8971
860	140	861	859
2000	8000	2001	1999

In the first problem **627** to find the nikhilam number we need to subtract last number 7 from 10 (10-7 =3). The remaining numbers 2 and 6 have to be subtracted from 9 (9-2=7, 9-6=3). So, the **Nikhilam number is 373.**

The **Ekhadhikena** of 627 is **628** and **Ekhanyunena** is **626.**

In the fifth problem **860** the last number is 0 hence, we should write 0 as it is. So, the number next to 0 is 6 which becomes the last number so subtract it from 10 (10-6= 4) and the remaining number is 8 subtract it from 9 (9-8= 1). So **Nikhilam number is 140.**

Ekhadhikena is **861** and **Ekhanyunena** is **859**.

Write Nikhilam, Ekhadhikena and Ekhanyunena for the following:

1. 20600
2. 348
3. 6789
4. 5823
5. 19
6. 1274
7. 919
8. 4600
9. 8964
10. 99

RIDDLE -2: I add 5 to 9 but get 2. The answer is correct. But how?

Chapter - 3

Base and Deviation

Base means 10 and powers of 10 like 10, 100, 1000, 10,000, 1,00,000 and so on. The base of the number is decided based on the nearest power of ten to that number.

Deviation is the difference between the number and base **(Number – Base)**.

Note: The number of digits in the deviation and the number of zeros in the base must always be equal.

If the number of digits in the deviation is less, then add zeros to make it equal.

For the numbers less than base we can apply Nikhilam sutra for easy subtraction.

If the deviation is negative, we write the negative sign on the top of the number and we call the number as **"REKHANK"**.

In Sanskrit **"REKH"** means bar and **"ANK"** means number so its collectively called as bar number.

Now let's look into some examples:

Find the base and deviation of the following numbers:

Numbers	Base	Deviation
16	10	16-10= + 6
7	10	7-10= $\overline{3}$
112	100	112-100 = + 12
92	100	92- 100= $\overline{08}$
1,017	1,000	1017-1000= + 017
994	1,000	994-1000= $\overline{006}$
10,009	10,000	10,009-10,000= + 0009
9985	10,000	9985-10,000= $\overline{0015}$

In the above example of **10,009** the nearest power of 10 for the number 10,009 is 10,000. So, the **base** is **10,000**.

To find the deviation we have to subtract (number – base) = (10,009-10,000= 9). But since there are 4 zeros in base (10,000) there has to be 4 digits in deviation. So as mentioned in the note to make it equal, we will add 3 zeros so the **deviation** becomes **+ 0009**.

In the above example of **994** the nearest power of 10 for the number 994 is 1000. So, the **base** is **1000**.

Since the number is less than base as mentioned in the note to find the deviation, we can apply Nikhilam sutra (10-4=6, 9-9=0 and 9-9=0). So, the **deviation** becomes $\overline{006}$.

Find the base and deviation for the following numbers:

1.	19	6.	995
2.	8	7.	10,018
3.	106	8.	9986
4.	87	9.	1,00,004
5.	1015	10.	99,998

RIDDLE – 3: A half is a third of it. What is it?

Chapter - 4

Vinculum Numbers

Vinculum Numbers are a group of both positive and negative numbers. There are few steps to be followed to convert a normal number to a vinculum number which are as follows:

Step – 1: Identify the numbers equal to (or) more than 5 and underline them.

Step – 2: Apply Nikhilam sutra to the underlined numbers and put a bar on the numbers to which the Nikhilam sutra is applied.

Step – 3: If the immediately preceding digit is less than 5 then, increase it by one (or) write its successor (Ekhadhikena).

Note: If there is no immediately preceding number then, write 0 and write its successor which is 1.

Let's understand it by some examples:

Normal Number	Vinculum Number
2876	$3\overline{124}$
98	$1\overline{02}$
3356711	$34\overline{43}311$
100672489	$101\overline{33}2\overline{511}$
37	$4\overline{3}$
4368	$44\overline{32}$

In the above example of **3356711** the digits 1 & 1 in the end are lesser than 5 so write 11 as it is.

Underline the numbers 567 because they are more than or equal to 5.

Now, apply Nikhilam sutra to the underlined numbers that is 567. So, last number 7 is subtracted from 10

(10-7=3) and remaining numbers 6 and 5 from 9

(9-6=3) and (9-5=4). So, the Nikhilam numbers are 433 and put a bar on it so it becomes $\overline{433}$.

The remaining preceding numbers are 3 and 3 which are less than 5. These numbers must not be treated individually instead, it has to be treated as a group. So, successor of 33 is 34. So, the answer is **34̄43311**.

In the next example of **98** both the numbers are more than 5 so underline it.

Apply Nikhilam to the underlined numbers so (10-8=2) and (9-9=0). So, it becomes 02 and put a bar on it which is $\overline{02}$.

There is no remaining preceding number left behind. So as mentioned in the note put 0 and write the successor of 0 which is 1. So, the answer is **1$\overline{02}$**.

Solve the following:

1. 2568	6. 9998
2. 4478	7. 379
3. 897	8. 19876
4. 112568	9. 981293
5. 30981065	10. 10029800

RIDDLE – 4: I am a 3-digit number. My tens digit is 6 more than unit's digit. My hundreds digit is 8 less than my tens digit. What number am I?

Chapter - 5

Conversion of Negative Number to Positive Number

There are few steps to be followed to bring a negative number to positive number which are given below:

Step-1: Identify the bar numbers and apply Nikhilam sutra to the bar numbers.

Step-2: If the remaining preceding number (number before bar number) is positive write its previous number (predecessor or Ekhanyunena).

Note: If 0 is present between two bar numbers then 0 is also treated to be negative that is we should write bar on the top of the 0.

Let us now understand it with few examples:

NEGATIVE NUMBER	POSITIVE NUMBER
$2\overline{12}$	188
$30\overline{412}$	29588
$\overline{4}3200\overline{5}40\overline{3}$	371995397
$40\overline{3}0\overline{3}$	39697
$3\overline{2}13\overline{1}$	28129

In the above problem $\overline{4}3200\overline{5}40\overline{3}$ the last number is a bar number so apply Nikhilam (10-3=7). The immediate number next to the bar number is 40. The previous number of 40 is 39.

We have another bar number which is $\overline{5}$. So, apply Nikhilam (10-5=5). The immediate number next to bar number is 200. So, the previous number of 200 is 199.

Now we have another bar number which is $\overline{3}$. So, apply Nikhilam (10-3=7). The immediate number next to bar number is 4. So, the previous number of 4 is 3.

So, the answer is **371995397**.

In the above problem of $40\overline{3}0\overline{3}$ we can see that 0 is present in between two negative (bar) numbers. So as mentioned in the note 0 is also treated to be negative so, the question becomes $40\overline{3}\overline{0}\overline{3}$.

Now apply Nikhilam sutra to the last three bar numbers (10-3=7), (9-0=9) and (9-3=6). The immediate number next to the bar number is 40. So, the previous number of 40 is 39.

So, the answer is **39697**.

Solve the following:

1. $1\overline{54}$
2. $10\overline{4}20\overline{5}$
3. $1\overline{3210}3$
4. $20\overline{34}$
5. $3\overline{2}13\overline{1}$

6. $4\overline{12}450\overline{2}$
7. $2\overline{11}$
8. $1\overline{1}$
9. $20\overline{2}00$
10. $235\overline{4}01\overline{3}$

RIDDLE – 5: How can you add eight 4's together so that the total adds up to 500?

Chapter - 6

Vedic Addition

In Vedic Mathematics addition can be performed in a very simple and fast manner by using a sutra called **Shuddha** method.

By using this method complex problems of addition can be solved in a very simple manner without using a calculator.

According to this method when we add two numbers if, the sum is more than 9 then, we keep a dot on the top of the second number which is called as the Shuddha dot.

Then we carry the unit (last) digit of the sum to the next number.

Let us now understand this by some examples:

Solve for the following:

1.   8+5+2+7+6

2.   56+86+71+68+12

3.   698+436+987+468+533

4.   7174+6378+7874+7929+5987

Now, first arrange these numbers one below the other.

8	5 6	6 •9 8	•7 1 •7 4
5•	•8 6•	•4 3 6•	6 3 7 8•
2	•7 1	•9 •8 7•	•7 •8 •7 4
7•	6 8•	4 6 8	•7 •9 2 9•
6	1 2	•5 • 3 3•	5 •9 •8 7•
28	**293**	**3122**	**35342**

In the above problem of **8+5+2+7+6** the first step is to write the numbers one below the other and add the first two numbers (8+5=13). 13 is more than 9.

So, keep a dot at 5 and carry the last digit of 13 which is 3 to the next number 2.

Then, (3+2=5). Then, add 5 to the next number 7

(5+7=12). 12 is again more than 9 so keep a dot at 7 and add the last digit of 12 that is 2 to the next number 6 (2+6=8).

Write 8 in the unit place of the answer.

Now, add the number of dots. The number of dots is 2. So, write 2 in the tens place. So, the answer is **28**.

In the next problem of **56+86+71+68+12** first write the numbers one below the other and then add the first two numbers of the last column (6+6=12).

12 is more than 9 so keep a dot to the second number 6 and then carry the unit digit of the number 12 which is 2 to the next number 1 (2+1=3).

Add 3 to the next number 8 (3+8=11). 11 is again more than 9 so keep a dot at 8 and then carry the unit digit of the number 11 which is 1 to the next number 2. So, (1+2=3). Write 3 in the unit digit of the answer.

Now, add the number of dots. There are 2 dots. So, add this to the next column number that is 5 (2+5=7).

Add 7 to the next number 8 (7+8=15). 15 is more than 9 so keep a dot at 8 and carry the unit digit of the number 15 which is 5 to the next number 7.

(5+7=12). 12 is again more than 9 so keep a dot at 7 and add the unit digit of the number 12 which is 2 to the next number 6.

(2+6=8). Now add 8 to the next number 1 (8+1=9). Now write 9 in the tens place.

Add the number of dots in this column. There are 2 dots. So, write 2 in the hundreds place. So, the answer is **293**.

Exercise sums:

1. 2+5+4+9+9

2. 8+4+9+5+2+4+8+1

3. 39+48+73+38

4. 582+693+844+358

5. 8776+3993+4576+5987

6. 89+80+75+21+18+65+29+36+50+39

7. 497+889+643+566

8. 7+6+8+1+9

9. 49+88+64+56+35+42+54+79+63+58

10. 65+26+36+50

RIDDLE - 6: Which is the only number from 0 to 1000 that has the letter **"a"** in it?

Chapter - 7

Multiplication Tables

The Multiplication Tables of any number of digits can be easily written by following the Shuddha method. We, know that multiplication is repeated addition. So, we can write the tables by adding the numbers. By using this method, we can save time and there is no need of memorizing the tables.

Let us understand it with few examples:

Write the tables of 13:

13 X 1 =	13
13 X 2 =	26
13 X 3=	39
13 X 4 =	52●
13 X 5 =	65
13 X 6 =	78
13 X 7 =	91●
13 X 8 =	104
13 X 9 =	117
13 X 10 =	130●

Note: The value of the dot is 1.

In the above problem first write the number given in the question that is 13 in the box. All the numbers are added with this 13 written in the box.

Now, write 13 in the first level because any number multiplied with 1 gives the same number.

Start adding from the last number that is, consider 3 in the box and the 3 in the first level as shown by an arrow (3+3=6) and write 6 in second level. Now add 6 with 3 in the box (6+3=9) and write 9 in third level.

Then add 9 with 3 in the box (9+3=12). 12 is more than 9 so write the unit digit of 12 which is 2 in the fourth level and keep a dot next to it.

Now add 2 with 3 (2+3=5) and write in fifth level.

Then add 5 with 3 (5+3=8) and write 8 in sixth level. Add 8 with 3 (8+3=11) which is more than 9 so write the unit digit of 11 which is 1 in seventh level by and keep a dot next to it.

Add 1 with 3 (1+3=4) and write 4 in eight level. Then add 4 with 3 (4+3=7) and write 7 in ninth level. Now add 7 with 3 (7+3=10) which is more than 9 so write unit digit of 10 which is 0 in tenth level and keep a dot next to it.

Now start adding the 1 in the box and the 1 in the first level (1+1=2) and write 2 in second level. Now, add 2 with 1 (2+1=3) and write 3 in third level.

Now add 3 with 1 (3+1=4) but there is a dot next to 2 so add the dot value which is 1 to 4 (4+1=5). So, write 5 in fourth level.

Add (5+1=6) and write 6 in fifth level. Add (6+1=7) and write 7 in sixth level. Add (7+1=8) but there is a dot so (8+1=9) and write 9 in seventh level.

Add (9+1=10) and write 10 in eight level. Add (10+1=11) and write 11 in ninth level. Add (11+1=12) but there is a dot so (12+1=13) write 13 in tenth level.

Write tables for 3999:

	3999
3999 X 1=	3 9 99
3999 X 2 =	7●9●98●
3999 X 3 =	11●9●97●
3999 X 4 =	15●9●96●
3999 X 5 =	19●9●95●
3999 X 6 =	23●9●94●
3999 X 7 =	27●9●93●
3999 X 8 =	31●9●92●

3999 X 9 =	35●9●91●
3999 X 10 =	39●9●90●

In the above problem first write 3999 in first level and start adding from the last number 9 in the box and 9 in the first level (9+9=18). 18 is more than 9 so keep a dot and write the unit digit of 18 which is 8 in the second level and keep a dot next to it.

Add 8 to 9 in the box (8+9=17). 17 is more than 9 so write the unit digit of 17 which is 7 in the third level and keep a dot next to it.

Add 7 to 9 in the box (7+9=16). 16 is more than 9 so write unit digit of 16 which is 6 in the fourth level and keep a dot next to it.

Add 6 to 9 in the box (6+9=15). 15 is more than 9 so write the unit digit of 15 which is 5 in the fifth level and keep a dot next to it.

Add 5 to 9 in the box (5+9=14). 14 is more than 9 so write the unit digit of 14 which is 4 in the sixth level and keep a dot next to it.

Add 4 to 9 in the box (4+9=13). 13 is more than 9 so write the unit digit of 13 which is 3 in the seventh level and keep a dot next to it.

Add 3 to 9 in the box (3+9=12). 12 is more than 9 so write the unit digit of 12 which is 2 in the eight level and keep a dot next to it.

Add 2 to 9 in the box (2+9=11). 11 is more than 9 so write the unit digit of 11 which is 1 in the ninth and keep a dot next to it.

Add 1 to 9 in the box (1+9=10). 10 is more than 9 so write the unit digit of 10 which is 0 in the tenth level and keep a dot next to it.

Add 9 in the tens place in the box to the 9 in the ten's place in the first level (9+9=18). Add 18 with the dot (18+1=19). 19 is more than 9 so write the unit digit of 19 which is 9 in all the levels that is till tenth level and keep a dot next to it. Because numbers are same it remains the same in all the levels.

Add 9 in the hundreds place in the box to the 9 in the hundred's place in the first level (9+9=18). Add 18 with the dot (18+1=19). 19 is more than 9 so write the unit digit of 19 which is 9 in all the levels that is till the tenth level and keep a dot next to it. Because the numbers are same it remains the same in all the levels.

Add 3 in the box to 3 in first level (3+3=6). Add 6 with the dot (6+1=7). Write 7 in second level.

Add 7 with 3 in the box (7+3=10). Add 10 with the dot (10+1=11). Write 11 in the third level. Since there is no immediately preceding number in the box, we can write 11 directly.

Add 11 with 3 in the box (11+3=14). Add 14 with the dot (14+1=15). Write 15 in fourth level.

Add 15 with 3 in the box (15+3=18). Add 18 with the dot (18+1=19). Write 19 in fifth level.

Add 19 with 3 in the box (19+3=22). Add 22 with the dot (22+1=23). Write 23 in the sixth level.

Add 23 with 3 in the box (23+3=26). Add 26 with the dot (26+1=27). Write 27 in the seventh level.

Add 27 with 3 in the box (27+3=30). Add 30 with the dot (30+1=31). Write 31 in the eight level.

Add 31 with 3 in the box (31+3=34). Add 34 with the dot (34+1=35). Write 35 in the ninth level.

Add 35 with 3 in the box (35+3=38). Add 38 with the dot (38+1=39). Write 39 in the tenth level.

Exercise problems:

Write tables for the following:

1. 12
2. 23
3. 303
4. 1034
5. 123412

6. 18
7. 78
8. 199
9. 1987
10. 18821

DID YOU KNOW? Every odd number has an "**e**" in it.

Chapter - 8

Nikhilam Multiplication

Nikhilam Multiplication is one of the easiest methods to find the product of 2 or more numbers. In this method both the multiplier and the multiplicand are compared to the nearest base.

Note: The number of digits in the deviation must always be equal to the number of zeros in the base.

If the number of digits in the deviation is less, then add zeros to make it equal.

If the number of digits in the deviation is more, then shift the extra digit to the other side.

In multiplication, the right-side of the answer is obtained by multiplying the deviations and the left side of the answer is obtained by adding or subtracting any one number with the other deviation. Adding or subtracting depends on the sign of the deviation.

SYMBOLS INDICATION:

B – BASE

D - DEVIATION

Numbers more than the base:

<u>**Multiply 12 X 13**</u>: <u>B=10</u>

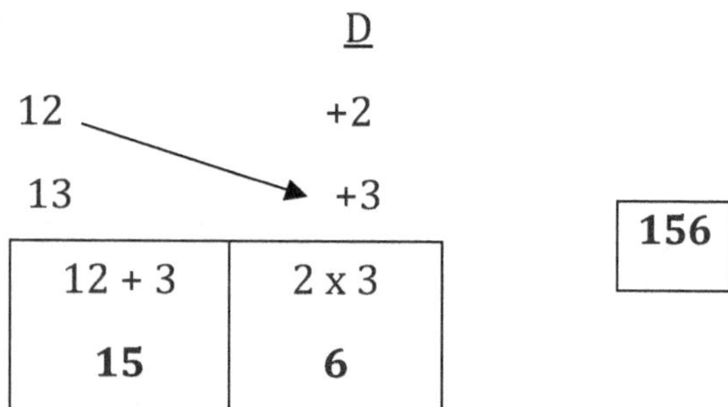

In the above problem **(12 X 13)** first write the numbers one below the other and identify the base nearest to both the numbers. So, the Base **(B)** is **10**.

Now identify the deviations for the numbers 12 and 13. So the Deviations **(D)** are **(+2 and +3)**.

Multiply the deviations (2 X 3=6) and write it on the right side. There is one zero in the base and there is one digit in the product of the deviation. Both are equal so its correct.

Now add one number with the other deviation

(12 + 3=15) or (13 + 2=15). You can consider any number with the other deviation. The answer remains the same, and write it on the left side.

Combine the answers on the left and the right side. So, the answer is **156**.

Numbers less than base:

Multiply 87 X 89: B = 100

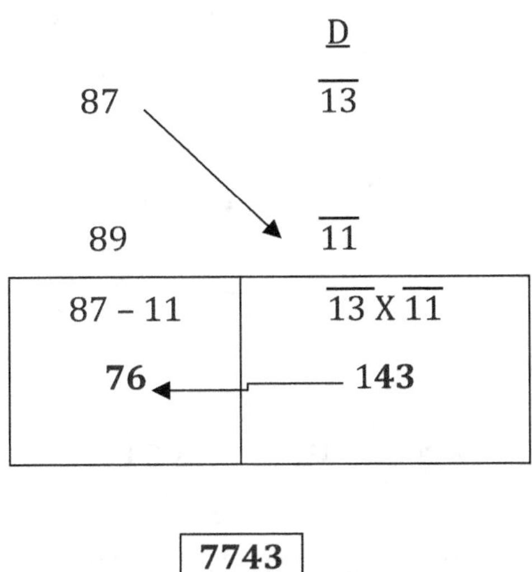

$$\boxed{7743}$$

In the above problem **(87 X 89)** first write the numbers one below the other and identify the base nearest to both the numbers. So, the Base **(B)** is **100**.

Now identify the deviations for the numbers 87 and 89. So the Deviations **(D)** are $(\overline{13} \text{ and } \overline{11})$.

Multiply the deviations $(\overline{13} \text{ X } \overline{11}=1\underline{43})$ and write it on the right side. When two negative numbers are multiplied answer is positive because **(- X - = +)**. There are two zeros in the base and there are three digits in the product of the deviation. There is one extra digit so keep the last two digits as it is and shift the extra digit to the left side.

Now subtract one number with the other deviation because the deviation is negative (87-11=76). Now add that extra digit 1 from the right side (76+1=77) and write it on the left side.

Combine the answers on the left and the right side. So, the answer is **7743**.

One number more and one number less than the base:

Multiply 991 X 1006:   B=1000

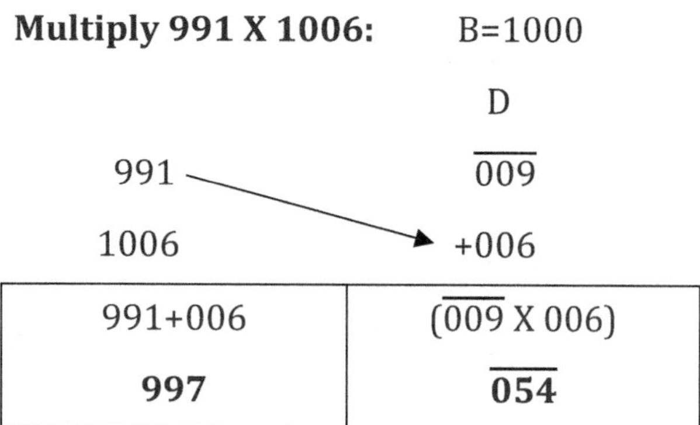

991+006	($\overline{009}$ X 006)
997	$\overline{054}$

997$\overline{054}$
996946

In the above problem **(991 X 1006)** first write the numbers one below the other and identify the base nearest to both the numbers. So, the Base **(B)** is **1000**.

Now identify the deviations for the numbers 991 and 1006. So, the Deviations **(D)** are **($\overline{009}$ and 006)**.

Multiply the deviations ($\overline{009}$ X 006= $\overline{054}$) and write it on the right side. When one negative number is multiplied with one positive number answer is negative because **(- X + = -)**. There are three zeros in the base and there are two digits in the product of the deviation. There is one digit less. So add one zero to make it equal.

Now, add one number with the other deviation

(991+ 006=997) and write it on the left side.

Combine the answers on the left and the right side. So, the answer is **997$\overline{054}$**.

We have negative numbers 054. So, apply Nikhilam sutra (10-4=6), (9-5=4), (9-0=9). The preceding number of 997 is 996. So, the final answer is **996946**.

Exercise problems:

1. 103 X 104
2. 15 X 17
3. 987 X 999
4. 992 X 1002
5. 95 X 105
6. 17 X 8
7. 991 X 1006
8. 996 X 994
9. 10,005 X 10,006
10. 111 X 109

RIDDLE – 7: I am an odd number, take away an alphabet and I become even. What number am I?

Chapter - 9

Squaring of Numbers Ending with 5

Squaring of numbers means multiplying the number by itself (or) multiplying the same number 2 times.

Squaring of numbers ending with 5:

Note: The square of 5 = (5^2) = (5 X 5 = 25).

On the right-hand side of the answer first write the square of 5 which is 25.

On the left-hand side of the answer multiply the remaining preceding number with its ekadhikena (successor).

Let us now understand this by few examples:

(15^2) =	225
(145^2) =	21025
(205^2) =	42025
(1035^2) =	1071225
$(99,9995^2)$ =	9,99,99,00,00,025

In the above problem of **(152)** first write the square of 5 which is 25 on the right side. Now, multiply the remaining number 1 with its ekadhikena (successor) 2 (1 X 2=2).

Now write 2 on the left side and combine both the values. So, the answer is **225**.

In the above problem of **(10352)** first write the square of 5 which is 25 on the right side. Now, multiply the remaining number 103 with its successor 104

(103 X 104=10712). You can solve this multiplication easily using Nikhilam Multiplication sutra explained in the previous chapter.

Now write 10712 on the left side and combine both the values. So, the answer is **1071225**.

In the above problem of **(99,99952)** first write the square of 5 which is 25 on the right side. Now, multiply the remaining number 99,9995 with its ekadhikena (successor) 1,00,000 (99,999 X 1,00,000 = 9,99,99,00,000).

Now write 9,99,99,00,000 on the left side and combine both the values. So, the answer is **9,99,99,00,00,025**.

Exercise problems:

1. (35^2)
2. (95^2)
3. (165^2)
4. (185^2)
5. (705^2)
6. (805^2)
7. (1065^2)
8. (1055^2)
9. (995^2)
10. (4005^2)

DID YOU KNOW: We follow the decimal system of numbers which has 10 digits from 0 to 9. It is also known as the Hindu Arabic Numeral System. It was discovered more than 10,000 years ago.

Chapter - 10

Squaring of Numbers Near to the Base

This is also called as Nikhilam squaring. This is one of the easiest ways to square the numbers provided the numbers are near to the base. By this method we can calculate the square of any number easily without taking much time.

Steps involved are as follows:

First find the base and deviation of the number.

Then find the square of the deviation and write it on the right side of the answer.

Add or subtract the number with the deviation and write it on the left side of the answer.

Now, combine the left and right parts of the answer.

Note: The number of digits in the square of the deviation must be equal to the number of zeros in the base.

If the number of digits is less than the zeros in the base, to make it equal add zeros.

If the number of digits is more than the zeros in the base, then shift the extra digit (carry over).

Let us understand by few examples:

Numbers more than the base:

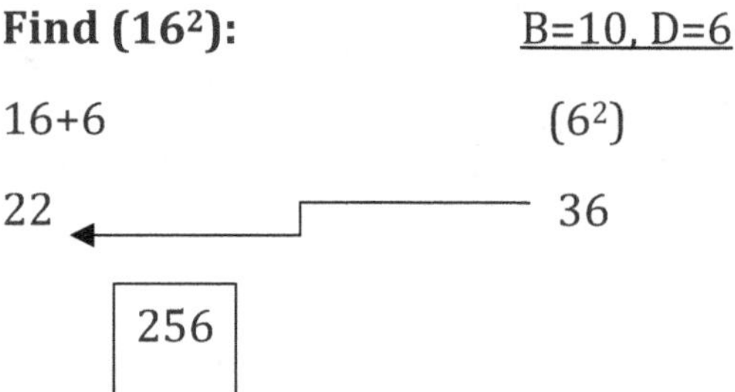

Find (16^2): $\quad$ B=10, D=6

16+6 $\quad$ (6^2)

22 $\quad\quad\quad\quad$ 36

256

In the above problem **(16^2)** first find the Base **(B)** of 16 which is **10**. Then find the Deviation **(D)** which is **6**.

Now find the square of 6 (6^2 = 6 X 6=3<u>6</u>). Write it on the right side. There is one zero in the base but there are two digits in the deviation which means there is an extra digit. So, keep the last digit 6 as it is and shift the extra digit 3 to the left side.

Now add the number 16 with the deviation 6 (16+6=22). Then (22+3=25). Write it on the left side.

Combine the answers from both the sides. So, the answer is **256**.

Find (1011²):   B=1000, D=011

1011+011   (011²)

1022   121

$$\boxed{1022121}$$

In the above problem **(1011²)** first find the Base **(B)** of 1011 which is **1000**. Then find the Deviation **(D)** which is **011**. There are 3 zeros in the base so we have to add 1 zero in the deviation to make it equal.

Now find the square of 011 ([011²] = 11 X 11=121). Write it on the right side. There are 3 zeros in the base and 3 digits in the deviation which is equal.

Now add the number 1011 with the deviation 011 (1011+011=1022). Write it on the left side.

Combine the answers from both the sides. So, the answer is **1022121**.

Numbers less than the base:

Note: For numbers less than the base to find the deviation apply Nikhilam sutra.

Let us now see few examples:

Find (9988²):   B=10,000, D=$\overline{0012}$

(9988-0012)   ($\overline{0012}$²)

9976   0144

$$\boxed{99760144}$$

In the above problem **(9988²)** first find the Base **(B)** of 9988 which is **10,000**. Then find the Deviation **(D)** which is $\overline{0012}$.

Now find the square of $\overline{0012}$ [$\overline{0012}$²] = $\overline{12}$ X $\overline{12}$=144. Write it on the right side. There are 4 zeros in the base and 3 digits in the deviation to make it equal we will add 1 zero. So, write 0144.

Now subtract the number 9988 with the deviation $\overline{0012}$ (9988-0012=9976). Write it on the left side.

Combine the answers from both the sides. So, the answer is **99760144**.

Find (997²):   B=100, D=$\overline{003}$

997-003   ($\overline{003}^2$)

994   009

994009

In the above problem **(997²)** first find the Base **(B)** of 997 which is **1000**. Then find the Deviation **(D)** which is $\overline{003}$.

Now find the square of $\overline{003}$ [$\overline{003}^2$] = $\overline{3}$ X $\overline{3}$ = 9. Write it on the right side. There are 3 zeros in the base and 1 digit in the deviation to make it equal we will add 2 zeros. So, write 009.

Now subtract the number 997 with the deviation $\overline{003}$ (997-003=994). Write it on the left side.

Combine the answers from both the sides. So, the answer is **994009**.

Exercise problems:

1. (142)   6.   (892)
2. (1082)   7.   (962)
3. (10122) 8.   (9932)
4. (10062) 9.   (99892)
5. (10,0092) 10. (99922)

DID YOU KNOW: Zero is the only number that cannot be represented in the Roman Numerals.

Chapter - 11

Multiplication With 11

Multiplication with 11:

Since 11 is a 2-digit number we should add 2 digits at a time.

Note: We should add 1 zero on either side of the number.

The number of digits in each column should be 1. If the number of digits is more than 1 then, it should be shifted to the next column.

Examples:

27 X 11 =	297
256 X 11 =	2816
9874 X 11 =	108614
674561 X 11 =	7420171
5648768 X 11 =	62136448

In the above problem **(256 X 11)** add 1 zero on either side of 256 as shown: **0 256 0**. Make sure you add zeros little far from the number.

We know that multiplication is repeated addition. So, start adding from the last number that is 0 by considering 2 digits at a time.

(0+6=6). Strike off 0. Now take next 2 numbers (6+5=11). Strike off 6. Then add (5+2=7). Strike off 5. Take next 2 numbers (2+0=2) = **2/7/11/6**.

But as mentioned in the note each column should contain only one digit. So, write 6 in the last. Next write 1 and shift 1 to 7 (1+7=8) so write 8. Then write 2. So, the answer is **2816**.

In the above problem **(5648768 X 11)** add 1 zero on either side of 5648768 as shown: **0 5648768 0**. Make sure you add zeros little far from the number.

We know that multiplication is repeated addition. So, start adding from the last number that is 0 by considering 2 digits at a time.

(0+8=8). Strike off 0. Now take next 2 numbers (8+6=14). Strike off 8. Then add (6+7=13). Strike off 6. Take next 2 numbers (7+8=15). Strike off 7. Add next 2 numbers (8+4=12). Strike off 8. Add next 2 numbers (4+6=10). Strike off 4. Add next 2 numbers (6+5=11). Strike off 6. Add next 2 numbers (5+0=5) = **5/11/10/12/15/13/14/8**.

But as mentioned in the note each column should contain only one digit. So, write 8 in the last. Next write 4 and shift 1 to 13 (1+13=14). Write 4 and shift 1 to 15 (1+15=16). Write 6 and shift 1 to 12 (1+12=13). Write 3 and shift 1 to 10 (1+10=11). Write 1 and shift 1 to 11 (1+11=12). Write 2 and shift 1 to 5 (1+5=6). Then write 6. So, the answer is **62136448**.

Solve the following:

1. 431 X 11
2. 25869 X 11
3. 543 X 11
4. 125123 X 11
5. 603 X 11
6. 214321 X 11
7. 86 X 11
8. 23 X 11
9. 54 X 11
10. 506 X 11

DID YOU KNOW: The sign Plus (+) and Minus (-) were discovered as early as 1489 A.D.

Chapter - 12

Vertical and Crosswise Method

This method works on the sutra **"Urdhva Tiryagbhyam"** which means vertical crosswise. This method is used for multiplication of numbers far from the base. In this chapter we will learn how to multiply a 2-digit number with any other 2-digit number in a simple way quickly.

Note: We must consider only 1 digit from each part. If the number of digits in each part is more than 1 then we must write the last digit as it is and shift the remaining digit.

Let us now consider few examples:

32 X 51 =	1632
24 X 33 =	792
32 X 41 =	1312
53 X 43 =	2279
23 X 52 =	1196

In the above problem of (24 X 33) first write the numbers one below the other as shown.

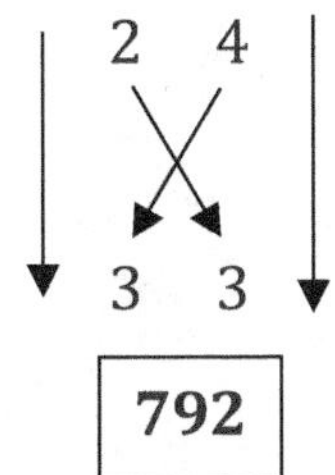

Now first multiply the last numbers that is 4 and 3

(4 X 3=12).

Then cross multiply the other two numbers and add those answers (2 X 3=6) and (4 X 3=12) = (12+6=18).

Now multiply the first numbers that is 2 and 3

(2 X 3=6) = **6/ 18/ 12.**

As mentioned in the note there has to be only 1 digit in each part. But we have 2 digits. So, let's start writing from the last number.

Write 2 in the unit digit of the answer as it is and shift 1 to 18 (1+18 =19) write 9 in tens place of the answer.

Shift 1 to 6 (1+6 =7) write 7 in hundreds place of the answer. So, the answer is **792**.

In the above problem of (53 X 43) first write the numbers one below the other as shown.

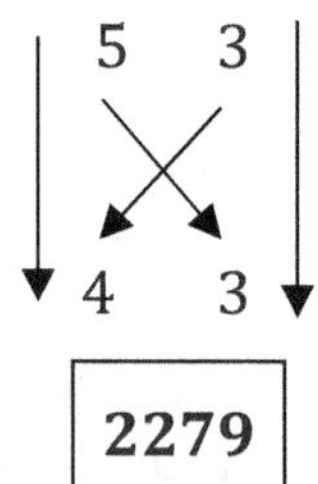

Now first multiply the last numbers that is 3 and 3

(3 X 3= 9).

Then cross multiply the other two numbers and add those answers (5 X 3=15) and (4 X 3=12) = (15+12=27).

Now multiply the first numbers that is 5 and 4

(5 X 4 =20) = **20/ 27/ 9.**

As mentioned in the note there has to be only 1 digit in each part. But we have 2 digits. So, let's start writing from the last number.

Write 9 in the unit digit of the answer as it is and then write 7 in tens place of the answer and shift 2 to 20 (2+20 =22) write 22. So the answer is **2279**.

Exercise problems:

1.	56 X 62	6.	63 X 21
2.	51 X 25	7.	34 X 85
3.	24 X 51	8.	85 X 42
4.	34 X 43	9.	41 X 58
5.	56 X 41	10.	53 X 49

RIDDLE – 8: What's the angle between the minute hand and hour hand at quarter past three?

Chapter - 13

Multiplication with 9'S

This method is used for multiplication of numbers with any number of 9's. There are 3 types of cases. **Multiplication when there are equal number of digits on left side and right side i.e. (multiplier and multiplicand).**

The steps are as follows:

First write the predecessor (one number before) of the multiplier and write this answer on the left side.

Then subtract the 9's with the predecessor and write this answer on the right side.

8 X 9 =	72
78 X 99 =	7722
639 X 999 =	638361
4782 X 9999 =	47815218
89165 X 99999 =	8916410835

Consider the example **(639 X 999)** here the number of digits on both the sides are equal.

First write the predecessor of the number 639 which is 638.

Now subtract 999 with the predecessor 638 (9-8=1),

(9-3=6) and (9-6=3) = 361.

Now combine the answers from both the sides. So, the answer is **638361.**

Consider the example **(89165 X 99999)** here the number of digits on both the sides are equal.

First write the predecessor of the number 89165 which is 89164.

Now subtract 99999 with the predecessor 89164

(9-4=5), (9-6=3), (9-1=8), (9-9=0) and (9-8=1) = 10835.

Now combine the answers from both the sides. So, the answer is **8916410835.**

DID YOU KNOW: The word "hundred" comes from the old Norse term, "hundrath", which actually means 120 and not 100.

When the numbers on multiplier is less than the numbers on multiplicand:

Note: When the numbers are less than the other side to make it equal add zeros.

The rest of the procedure remains the same.

Examples:

83 X 9999 =	829917
46 X 999 =	45954
8 X 99 =	792
456 X 999999 =	455999544
487 X 9999 =	4869513

Consider the example **(83 X 9999)** here the number of digits on both the sides are not equal. The number of digits in the multiplier is 2 and multplicand is 4. So, to make it equal we will add 2 zeros before the multiplier and rewrite the question as **(0083 X 9999).**

First write the predecessor of the number 0083 which is 0082.

Now subtract 9999 with the predecessor 0082 (9-2=7),

(9-8=1), (9-0=9) and (9-0=9) = 9917.

Now combine the answers from both the sides. So, the answer is **829917.**

Consider the example **(456 X 999999)** here the number of digits on both the sides are not equal. The number of digits in the multiplier is 3 and the multiplicand is 6. So, to make it equal we will add 3 zeros before the multiplier and rewrite the question as **(000456 X 999999).**

First write the predecessor of the number 000456 which is 000455.

Now subtract 999999 with the predecessor 000455

(9-5=4), (9-5=4), (9-4=5), (9-0=9), (9-0=9) and (9-0=9) = 999544.

Now combine the answers from both the sides. So, the answer is **455999544.**

DID YOU KNOW? A geometrical close bounded figure which has 20 sides is known as an "Icosagon".

Multiplication when number of digits on multiplicand is less than the multiplier:

Note: The negative number that we get on the right side has to be shifted to left side and then simplified to get the answer.

Examples:

5684 X 99 =	562716
34564 X 99 =	3421836
21099 X 999 =	21077901
35999 X 99 =	3563901
6417 X 99 =	635283

Consider the example **(35999 X 99)** here the number of digits on both the sides are not equal. The number of digits on the multipier is 5 and the multiplicand is 2. So, to make it equal we will add 3 zeros before the multiplicand and rewrite the question as **(35999 X 00099).**

First write the predecessor of the number 35999 which is 35998.

Now subtract 00099 with the predecessor 35998

(9-8=1), (9-9=0), (0-9=$\bar{9}$), (0-5=$\bar{5}$) and (0-3=$\bar{3}$) = $\overline{359}01$ = 35998/ $\overline{359}01$.

Now this negative numbers 359 have to be shifted to left side and subtracted = (35998-359 = 35639)

Now combine the answers from both the sides. So, the answer is **3563901.**

Consider the example **(6417 X 99)** here the number of digits on both the sides are not equal. The number of digits on the multiplier is 4 and multiplicand is 2. So, to make it equal we will add 2 zeros before the multiplicand and rewrite the question as **(6417 X 0099).**

First write the predecessor of the number 6417 which is 6416.

Now subtract 0099 with the predecessor 6416

(9-6=3), (9-1=8), (0-4=$\bar{4}$) and (0-6=$\bar{6}$) = $\overline{64}83$ =

6416/ $\overline{64}83$.

Now this negative numbers $\overline{64}$ has to be shifted to left side and subtracted = (6416-64 = 6352)

Now combine the answers from both the sides. So, the answer is **635283.**

Examples:

1. 686 X 999
2. 3456 X 9999
3. 124345 X 99999
4. 789 X 99999
5. 726 X 9999
6. 6188 X 999999
7. 30001 X 99
8. 213 X 99
9. 102030 X 99
10. 381347 X 99

RIDDLE – 9: If 1+9+8=1, then find the value of 2+8+9.

Chapter - 14

Multiplication of Numbers When Sum of Last Digits is 10, 100 and 1000.

When the sum of last few digits of multiplier and multiplicand is:

Sum of numbers	LHS	RHS
10	–	2 digits
100	0	3 digits
1000	00	4 digits

Multiplication when sum of last digits is 10:

Note: The sum of last digits of the multiplier and the multiplicand must be 10 provided the previous digits are same.

Multiply the last digits which add up to 10, 100 or 1000.

Then multiply the previous digits with its successor and then compare it with the table given above and write the answer.

Examples:

43 X 47 =	2021
122 X 128 =	15616
61 X69 =	4209
704 X 706 =	497024

In the above problem of **(43 X 47)** observe the unit's digits of the both multiplier and multiplicand that is 7 and 3. When we add both the digits we get (7+3=10).

The previous numbers must be same. So, first multiply (7 X 3 =21) and write 21 on right side. Compare it with the table.

RHS should have 2 digits and there are 2 digits in RHS so, write 21 on right side.

Then multiply the previous digit 4 with its successor 5 (4 X 5 = 20) and write 20 on left side.

So, the final answer is **2021**.

In the above problem of **(704 X 706)** observe the unit's digits of the both multiplier and multiplicand that is 6 and 4. When we add both the digits we get (6 + 4 = 10).

The previous numbers must be same. So, first multiply (6 X 4 = 24). Compare it with the table. RHS should have 2 digits. So, write 24 on right side

Then multiply the previous digit 70 with its successor 71 (70 X 71 = 4970) write 4970 on let side.

So, the final answer is **497024**.

DID YOU KNOW? The square root of 2 is 1.41 which was the first irrational number to be discovered which is known as the **"PHYTAGORAS CONSTANAT"**.

Multiplication when sum of last digits is 100:

Note: Even if there is a 0 in LHS you have to add another 0 according to the above given table.

Examples:

496 X 404 =	200384
768 X 732 =	562176
9099 X 9001 =	81900099
108 X 192 =	20736

In the above problem of **(496 X 404)** observe the last 2 digits of both the multiplier and multiplicand that is 96 and 04. When we add both the numbers we get (96+04=100).

The previous numbers must be same. So, first multiply (96 X 04=384) and write 384 on right side. Compare it with the table.

RHS should have 3 digits and there are 3 digits in the RHS so write 384.

Then multiply the previous digit 4 with its successor 5 (4 X 5 = 20) according to the table we should add 1 zero so write 200 on LHS.

So, the final answer is **200384**.

In the above problem of **(768 X 732)** observe the last 2 digits of both the multiplier and multiplicand that is 68 and 32 when we add both the numbers we get (68+32=100).

The previous numbers must be same. So, first multiply (68 X 32=2176) you can use the sutra of vertical crosswise method and write 2176 on right side. Compare it with the table.

RHS should have 3 digits but there are 4 digits in the RHS. So, we should keep the last 3 digits as it is and shift the extra digit to LHS so write 176 and shift 2.

Then multiply the previous digit 7 with its successor 8 (7 X 8=56) according to the table we should add 1 zero so write 560 on LHS. Then we should add 560 with the extra digit 2 (560+2=562) so write 562 on LHS.

So, the final answer is **562176**.

DID YOU KNOW? Pi is also known as the ratio of circumference to diameter of a circle. It is a special number which is irrational. There is a designated day called Pi Day that we celebrate on March 14.

Multiplication when sum of last digits is 1000:

Note: Even if there is a 0 in LHS you have to add 00 according to the above given table.

Examples:

5989 X 5011 =	300010879
9999 X 9001 =	90000999
6002 X 6998 =	42001996
4992 X 4008 =	20007936

In the above problem of **(9999 X 9001)** observe the last 3 digits of both the multiplier and multiplicand that is 999 and 001 when we add both the digits we get (999+001=1000).

The previous numbers must be same. So, first multiply (999 X 001=999) and write 999 on right side. Compare it with the table.

RHS should have 4 digits but there are 3 digits in the RHS. So, add 1 zero and write 0999.

Then multiply the previous digit 9 with its successor 10 (10 X 9=90) according to the table we should add 2 zeros so write 9000 on LHS. So, the final answer is **90000999**.

In the above problem of **(5989 X 5011)** observe the last 3 digits of both the multiplier and multiplicand that is 989 and 011 when we add both the digits we get (989+011=1000).

The previous numbers must be same. So, first multiply (989 X 011=10879) and write 10879 on right side. To achieve this quickly, apply the multiplication with 11 sutra. Compare it with the table.

RHS should have 4 digits but there are 5 digits in the RHS. So, we should keep the last 4 digits 0879 as it is and shift the extra digit 1 to LHS.

Then multiply the previous digit 5 with its successor 6 (5 X 6=30) according to the table we should add 00 zeros so write 3000 on LHS. Then we should add 3000 with the extra digit 1 (3000+1=3001) so write 3001 on LHS.

So, the final answer is **30010879**.

Solve the following:

1. 124 X 126
2. 34 X 36
3. 309 X 301
4. 111 X 189
5. 4088 X 4012
6. 2011 X 2099
7. 4988 X 4012
8. 10,994 X 10,006
9. 8999 X 8001
10. 7980 X 7020

RIDDLE – 10: If there are 70 students in your class and you shake every student's hand when you arrive, how many handshakes will you have to make?

Model Paper

NOTE: Answer all the questions

Write all the steps in detail.

Total marks = 100; Total time = 2hr.

I. Find out the Beejank for the following numbers: (2 X 2=4)

1. 54126729
2. 34156892

II. Write Nikhilam, Ekhadhikena and Ekhanyunena for the following numbers: (2 X 6=12)

1. 9889
2. 92907

III. Convert the following numbers into Vinculum numbers: (2 X 2=4)

1. 698
2. 46157

IV. Convert the following Vinculum numbers into normal numbers: (2 X 2=4)

1. $1\overline{34}$
2. $6\overline{2}1$

V. Multiply the following: (12 X 3=36)

1. 108 X 109
2. 85 X 93
3. 115 X 96
4. 83 X 87
5. 704 X 796
6. 42 X 35
7. 74 X 86
8. 836 X 999
9. 6432 X 999
10. 47 X 999

11. 3752 X 11

12. 4305 X 11

VI. Square the following numbers: (4 X 3=12)

1. 94^2

2. 1018^2

3. 65^2

4. 85^2

VII. Write the base and deviation for the following numbers: (2 X 2=4)

1. 9993

2. 1019

VIII. Write the multiplication tables: (2 X 4=8)

1. 44

2. 189

IX. Solve the following: (6 X 2=12)

1. 3 X 6 X 5 X 6

2. 1+ (2+4) + 6+ 5

3. 70 + (6/3)

4. 3- 9- 2+6

5. 75/5

6. 18 X 5 X 0 X 8

X. Add the following using Shuddha method:

(2 X 2=4)

1. 99+47+16+35

2. 645+283+394+785

Riddle Answers

RIDDLE – 1:   The other name for a circle is Perigon. It means an angle of 360 **degree.**

RIDDLE – 2:   When it is 9 a.m., add 5 hours to it and you will get 2.p.m.

RIDDLE – 3:   It is **1½.**

RIDDLE – 4:   The number is **193.**

RIDDLE – 5:   **4+4+4+44+444=500**

RIDDLE – 6:   **"One Thousand".**

RIDDLE – 7:   **"(SEVEN – S=EVEN)"**

RIDDLE – 8:   The angle is **7.5 degree.**

RIDDLE – 9:   1(One) + 9(Nine) +8(Eight) = **ONE =1.** So, 2(Two) + 8(Eight) + 9(Nine) = **TEN =10.**

RIDDLE – 10:   **69** handshakes.

Answers

CHAPTER - 1

Beejank:

1) 5		6) 1
2) 2		7) 6
3) 9		8) 9
4) 6		9) 3
5) 8		10) 8

CHAPTER - 2

Nikhilam, Ekhadhikena and Ekanyunena Purvena sutras:

1) 79400, 20601, 20599
2) 652, 349, 347
3) 3211, 6790, 6788
4) 4177, 5824, 5822
5) 81, 20, 18
6) 8726, 1275, 1273
7) 081, 920, 918
8) 5400, 4601, 4599
9) 1036, 8965, 8963
10) 01, 100, 98

CHAPTER - 3

Base and Deviation:

1) 10, +9

2) 10, $\overline{2}$

3) 100, +06

4) 100, $\overline{13}$

5) 1000, +015

6) 1000, $\overline{005}$

7) 10,000, +0018

8) 10,000, $\overline{0014}$

9) 1,00,000, +00004

10) 1,00,000, $\overline{00002}$

CHAPTER - 4

Vinculum numbers:

1) $3\overline{432}$

2) $45\overline{22}$

3) $1\overline{103}$

4) $113\overline{432}$

5) $31\overline{021}1\overline{35}$

6) $1\overline{0002}$

7) $4\overline{21}$

8) $2\overline{0124}$

9) $1\overline{0}213\overline{1}3$

10) $1003\overline{0}2\overline{0}0$

CHAPTER - 5

Conversion of negative number to positive number:

1) 046

2) 96195

3) 067897

4) 1966

5) 28129

6) 3876498

7) 189

8) 09

9) 19800

10) 2346007

CHAPTER - 6

Vedic addition:

1) 29	6) 502
2) 41	7) 2595
3) 198	8) 31
4) 2477	9) 588
5) 23,332	10) 177

CHAPTER - 7

Multiplication tables:

1)   12, 24, 36, 48, 60, 72, 84, 96, 108, 120.

2)   23, 46, 69, 92, 115, 138, 161, 184, 207, 230.

3)   303, 606, 909, 1212, 1515, 1818, 2121, 2424, 2727, 3030.

4)   1034, 2068, 3102, 4136, 5170, 6204, 7238, 8272, 9306, 10,340.

5)   1,23,412, 2,46,824, 3,70,236, 4,93,648, 6,17,060, 7,40,472, 8,63,884, 9,87,296, 11,10,708, 12,34,120.

6)   18, 36, 54, 72, 90, 108, 126, 144, 162, 180.

7)   78, 156, 234, 312, 390, 468, 546, 624, 702, 780.

8)   199, 398, 597, 796, 995, 1194, 1393, 1592, 1791, 1990.

9)   1987, 3974, 5961, 7948, 9935, 11,922, 13,909, 15,896, 17,883, 19,870.

10) 18,821, 37,642, 56,463, 75,284, 94,105, 1,12,926, 1,31,747, 1,50,568, 1,69,389, 1,88,210.

CHAPTER – 8

Nikhilam Multiplication:

1) 10,712

2) 255

3) 9,86,013

4) 9,93,984

5) 9,975

6) 136

7) 9,96,946

8) 9,90,024

9) 10,01,10,030

10) 12,099

CHAPTER - 9

Squaring of numbers ending with 5:

1) 1225

2) 9025

3) 27,225

4) 34,225

5) 4,97,025

6) 6,48,025

7) 11,34,225

8) 11,13,025

9) 9,90,025

10) 1,60,40,025

CHAPTER - 10

Squaring of numbers near to the base:

1) 196

2) 11,664

3) 10,24,144

4) 10,12,036

5) 10,01,80,081

6) 7921

7) 9216

8) 9,86,049

9) 9,97,80,121

10) 9,98,40,064

CHAPTER -11

Multiplication with 11:

1) 4741	6) 23,57,531
2) 2,84,559	7) 946
3) 5973	8) 253
4) 13,76,353	9) 594
5) 6633	10) 5566

CHAPTER - 12

Vertical crosswise method:

1) 3472	6) 1323
2) 1275	7) 2890
3) 1224	8) 3570
4) 1462	9) 2378
5) 2296	10) 2597

CHAPTER - 13

Multiplication with 9's:

1) 6,85,314	6) 6,18,79,93,812
2) 3,45,56,544	7) 29,70,099
3) 12,43,43,75,655	8) 21,087
4) 7,88,99,211	9) 1,01,00,970
5) 72,59,274	10) 3,77,53,353

CHAPTER - 14

Multiplication of numbers when the sum of last digits is 10, 100 and 1000:

1) 15,624

2) 1224

3) 93,009

4) 20,979

5) 1,64,01,056

6) 42,21,089

7) 2,00,11,856

8) 11,00,05,964

9) 7,20,00,999

10) 5,60,19,600

From the Author

I appreciate your interest in learning Vedic Mathematics.

My sole intention of writing this book is to impart knowledge in the field of Vedic Mathematics as I understand that it is useful in more ways than one.

I hope you found this book useful. I am working on 3 more books as a follow up to this book.

I look forward to you purchasing and working on those as well.

This book and the rest are and will be written in a very simple manner such that everybody will be able to understand and absorb the knowledge and wisdom of the Vedas contained in them.

My gratitude to everyone reading this book and my sincere request to you to share the knowledge with as many people as you can.

Thanks and Best Regards,

Akshatha Amruth